FORGOTTEN HISTORY OF EVERYDAY INVENTIONS

by Patricia Hutchison

12 STORY LIBRARY

www.12StoryLibrary.com

12-Story Library is an imprint of Bookstaves.

Photographs ©: David Duprey/Associated Press, cover, 1; Brett Behrens/Associated Press, 4; Marco Consani/CC3.0, 5; PD, 6; PD, 7; Daderot/CC0, 8; Bettmann/Getty Images, 9; Bill Buxton/CC4.0, 10; Edd Thomas/CC2.0, 10; Todd Ehlers/CC2.0, 11; PD, 12; Monkey Business Images/Shutterstock.com, 13; SRI International/CC3.0, 14; Christopher W./CC2.0, 15; Tinkeringbell/PD, 16; Rawpixel.com/Shutterstock.com, 17; PD, 18; Hopefulromntic/CC3.0, 19; Jong42/Shutterstock.com, 20; Josep Curto/Shutterstock.com, 20; NASA, 21; United States Patent Office/PD, 22; PD, 23; testing/Shutterstock.com, 23; SenseiAlan/CC2.0, 24; Dwight Burdette/CC3.0, 25; wilaiwan jantra/Shutterstock.com, 25; INTERFOTO/Alamy Stock Photo, 26; Chronicle/Alamy Stock Photo, 26; Keith Homan/Shutterstock.com, 27; Jari Hindstroem/Shutterstock.com, 28; Everett Historical/Shutterstock.com, 29

ISBN
978-1-63235-592-8 (hardcover)
978-1-63235-646-8 (paperback)
978-1-63235-705-2 (ebook)

Library of Congress Control Number: 2018947714

Printed in the United States of America
Mankato, MN
July 2018

About the Cover

A Frisbee and an original Frisbie's pie tin at the National Museum of Play in Rochester, New York.

12 STORY LIBRARY

Access free, up-to-date content on this topic plus a full digital version of this book. Scan the QR code on page 31 or use your school's login at 12StoryLibrary.com.

Table of Contents

1

Pie Tins Inspire a Favorite Toy

Inventor Ed Headrick in 1985.

In the 1940s, college students in New England loved Frisbie's pies. The fresh fruit baked between two flaky crusts tasted delicious. The Frisbie Pie Company sold thousands every day. Each pie came in a round tin. After eating the pies, students would throw the empty tins back and forth. The thrower called out "Frisbie!" to warn the catcher. The term became the name of a new sport. Professors at Princeton University thought the game was a passing fad.

In 1948, business partners Walter Frederick Morrison and Warren Franscioni invented their own flying disc. This disc was made of plastic. It flew farther than pie tins. It also allowed for better aim. Seven years later, Morrison made improvements to the disc. He sold the idea to a new toy company, Wham-O. Wham-O called the disc the Pluto Platter.

They hoped the space-themed name would attract buyers. A year later, Wham-O changed the name to Frisbee.

Ed Headrick was an inventor at Wham-O. He received a patent in 1967 for the modern Frisbee. The improved disc had raised rings on its surface. The rings gave the disc better stability while in the air. By 1977, Wham-O had sold more than 100 million Frisbees. Today dozens of companies sell flying discs. Several sports use the discs in competitions. Worldwide, Frisbee has been more than a passing fad.

200 million

Approximate number of Frisbees sold.

- In the 1940s, college students used pie tins as flying discs.
- Inventors made a plastic version of the flying disc in 1948.
- The toy company Wham-O sold the first Frisbee.
- Today flying discs are used in a variety of sports.

2

Basketball Begins in a Blizzard

In December 1891, a blizzard hit Springfield, Massachusetts. The young men at the YMCA training school became restless. They longed to play sports outside.

Their instructor, James Naismith, was trying to come up with a new sport. It had to include large numbers of men. It would have to be played indoors. The sport couldn't be rough, but it had to give the men plenty of exercise.

Naismith went to the janitor. He asked for two boxes to use for goals. The janitor brought him two peach baskets. Naismith nailed them at each end of the gym, ten feet above the floor. Next Naismith posted the rules for the game.

Two teams of nine players ran onto the court. Naismith tossed a soccer ball in the air, and basketball was born. William Chase sank the first basket. The game ended with a score of 1-0. That game was quite rough, however. Men kicked and tackled. Several were injured. Naismith made up more rules. Players weren't allowed to run while carrying the ball. This stopped them from tackling.

Word of the new sport spread quickly. The first professional game was

Dr. James Naismith with a ball and peach basket.

THINK ABOUT IT

In 1900, most Americans earned about $8 a week. So $15 was the equivalent of two weeks' pay. How does that compare to what professional basketball players earn today?

played in 1896. Each man was paid fifteen dollars. In 1905, basketball became a permanent winter sport. In 1936, it was added to the Olympics. Today it's played all over the world, and not only by men. Women have played basketball since 1892.

13

Number of original rules for basketball.

- James Naismith invented basketball in 1891.
- The first game was played with a soccer ball and two peach baskets.
- The first professional game was played in 1896.
- In 1936, basketball was added to the Olympics.

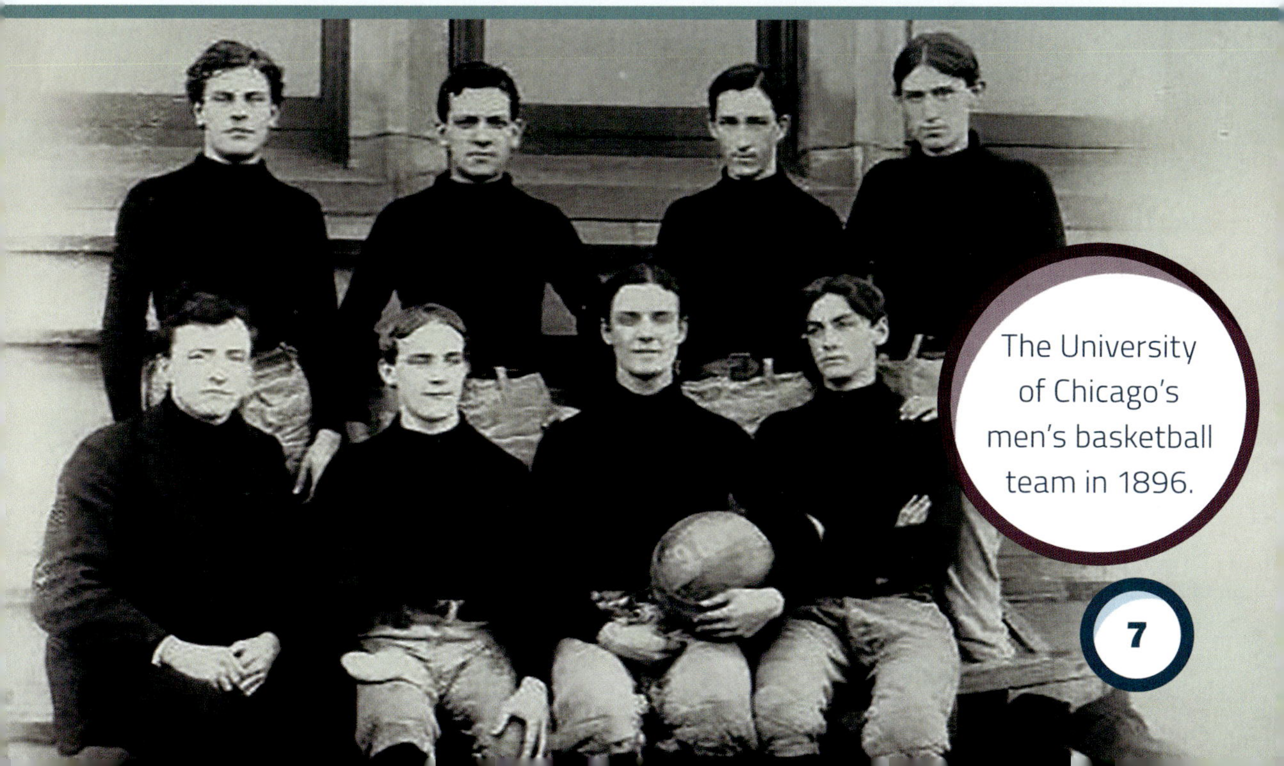

The University of Chicago's men's basketball team in 1896.

3 Microwaves Can Cook

Magnetrons are vacuum tubes that release microwave energy. During World War II, they were used in radar systems to detect planes. One day in 1946, Percy Spencer was testing magnetrons. He stuck his hand in his pocket and felt a gooey mess. His peanut cluster bar had melted. Spencer wondered how that happened. Deciding to experiment, he placed a raw egg next to the magnetron. The egg began to vibrate. Then it exploded. The next day he brought in some dried corn. He set it near the magnetron and watched it pop. Spencer realized that the microwave energy had cooked the food.

A World War II era magnetron.

Spencer built a metal box. He sent the microwave power to it. The energy couldn't escape through the metal. The microwave oven was born. Spencer filed a patent for his invention in 1945.

Two years later, the Raytheon Company introduced the first commercial microwave oven. It was five feet tall and weighed 750 pounds. It cost $5,000. The oven's popularity grew, mostly in restaurants. Chefs could heat food quickly in the microwave ovens. Customers enjoyed the fast service.

Microwave technology improved slowly. In 1967, the Amana Radar range was sold for household use. It was small enough to fit on a kitchen counter and cost about $500. Microwave ovens soon became more popular than dishwashers. In 1975, only four percent of homes in the United States had the appliance. Today almost everyone owns one.

A model demonstrating the Raytheon Radarange's cooking abilities.

90

Percent of households in the US that have a microwave oven.

- In 1946, Percy Spencer discovered that microwave radiation could cook food.
- Spencer patented the first microwave oven in 1945.
- The first commercial microwave oven weighed 750 pounds.
- In 1967, a household microwave oven was introduced.

THINK ABOUT IT

Find out how microwave ovens cook food. Why do they cook much faster than a regular oven?

4 Couch Potatoes Are Born

Years ago, TV watchers had to get up from their chairs to change channels on their sets. In 1950, Zenith introduced the first TV remote. It was called the Lazy Bones. People were finally able to control their TVs without leaving their seats. There was one problem, however. People often tripped over the cord that attached the remote to the TV.

The founder of Zenith demanded a better product. In 1955, Eugene Polley invented the Flash-Matic. It was the first wireless remote control. The device looked like a gun. It worked by "shooting" light signals at the TV. Receptors in the four corners of the screen received the signals.

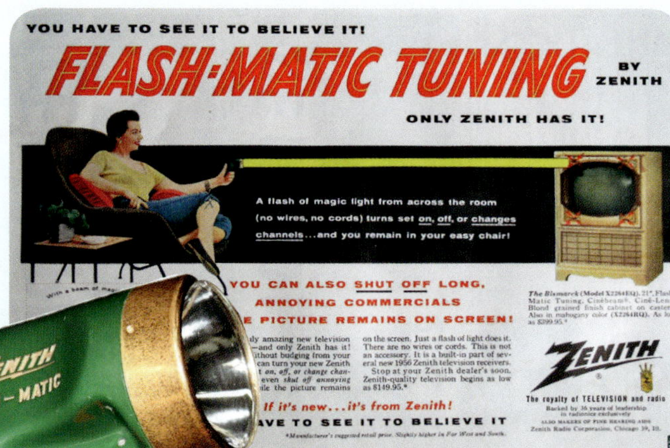

A Zenith Lazy Bones TV remote and a Flash-Matic (above).

MEN OF THE MILLENNIUM

Robert Adler is considered the father of the remote control. This made Eugene Polley somewhat bitter. But in 1997, both men received an Emmy Award for their inventions. In 1999, *Sports Illustrated* named them its Men of the Millennium.

100

Dollars a remote control added to the cost of a TV set in 1955.

- Zenith developed a wired TV remote in 1950.
- Eugene Polley invented the first wireless remote using light signals.
- Robert Adler used sound technology for his TV remote.
- Today's TV remotes use infrared signals.

The bottom receptors turned the TV on or off. The top ones changed the channels. The Flash-Matic was expensive. It also had some technical problems. Sometimes the TV channels would change if bright sunlight hit the receptors.

Robert Adler solved that problem in 1956. He invented the Space Command. The device used tiny hammers that struck metal rods. This sent commands by sound waves to the TV. The Space Command also had some faults. The sound of jingling keys could send commands to the TV. Even so, these remotes were used for over 25 years. Today's technology is more like Polley's. Remotes now use infrared light to control the TV.

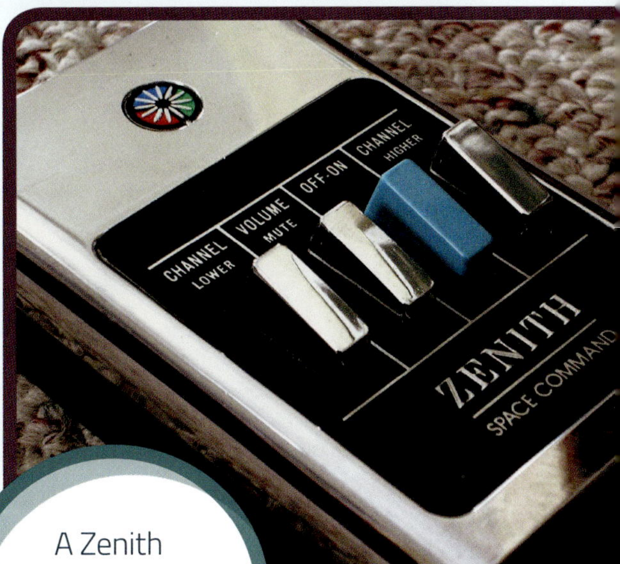

A Zenith 1956 Space Command device.

5 Carts Make Shopping Easier

Years ago, people shopped at the local general store. Customers waited at the counter while clerks gathered their orders. Then in 1916, the first self-serve grocery store opened in Tennessee. This new way of shopping spread quickly throughout the United States.

In 1934, Sylvan Goldman opened a Humpty Dumpty supermarket in Oklahoma. He watched as his customers filled wire baskets with groceries. They had to lug the heavy items to the checkout stand. Sylvan knew there had to be a better way to shop.

Sitting in his office one day, Goldman was inspired by a folding chair. He put a wire basket on top of the chair. Then he added castors on the legs. Goldman invented the first shopping cart. People could now push the cart around, piling their groceries in it.

Men were insulted. They felt their arms were strong enough to carry the groceries. The carts reminded women of baby carriages. They refused to use them. After a while, people realized how helpful the carts could be. Today people use them in nearly every store. Next to the automobile, shopping carts are the most popular vehicle in the world.

25 million

Number of shopping carts in use worldwide.

- The first supermarket opened in 1916.
- People used wire baskets to carry their groceries.
- In 1937, Sylvan Goldman invented the shopping cart.
- At first, people didn't like using shopping carts.

THINK ABOUT IT

Think about a chore that you don't like doing. What is something you might invent to help you do it more easily? Draw a picture or write a description of your invention.

6

Just Point and Click

Computers weren't user-friendly in their early days. Technicians typed commands or punched cards to make them work. Then, in 1963, Douglas Engelbart came up with an invention. He described it as an "X-Y position indicator for a display system." It was housed in a large wooden box. It moved on two wheels. Buttons on top gave commands to the computer.

When the device was introduced in 1981, it was called a mouse. It was part of a personal computer made by Xerox. Users no longer had to type commands. They used the mouse to click icons on the screen. However, few people bought this computer.

In 1979, Xerox allowed Steve Jobs to see the mouse. Jobs had helped start Apple Computer. He thought the mouse was amazing. He told his designers at Apple to build a better mouse. It had to be usable on any surface. And it had to cost less than 15 dollars to make.

In 1984, Apple released the first successful computer run by a mouse. Personal and commercial computers continued to use

A prototype of an early computer mouse.

text-based commands. In 1995, Microsoft introduced the Windows 95 operating system. It also used a mouse. Today many computers use touch screens. Some experts think the computer mouse will soon become extinct.

3

Number of buttons on the first computer mouse.

- The first computers used typed commands and punch cards.
- In 1963, Douglas Engelbart invented the first computer mouse.
- Apple released the first successful mouse-based computer in 1984.
- The Windows 95 operating system, introduced in 1995, used a mouse.

MICE ARE NICE

Today's mice are smaller and easier to handle. They use lasers instead of wheels. But they still work the same way. The computer records the distance and speed the mouse travels. It turns that information into code. Then it displays the information on the screen.

7 Sticky Notes Are a New Way to Communicate

In 1968, Spencer Silver was trying to develop a stronger type of glue. It was part of his job as a chemist at the 3M company. What he created were adhesive microspheres. But they were not what he was hoping for. The new glue kept its stickiness but could be easily removed.

Inventor Arthur Fry with his Post-it Note.

Spencer tried for years to find a use for microspheres.

Six years later, a coworker named Art Fry contacted Spencer. Fry had an idea. He wanted a bookmark that would stick to a page but could be easily removed. Fry thought Spencer's microspheres would work perfectly. He made a prototype. Fry started writing on the bookmarks, making labels out of them. Soon his office team was writing messages on the yellow squares. The Post-it Note was born. The sticky notes became a new way to communicate.

3M launched the product in 1977, but it failed. Consumers wouldn't try them. A year later, the company gave out free samples. Ninety percent of the people said they would buy them. By 1980, Post-it Notes were sold all over the United States. The next year, 3M launched them in Canada and Europe. Today the colorful notes are used in homes and offices everywhere.

11

Average number of messages on Post-it Notes a professional receives each day.

- Spencer Silver invented microspheres, a new type of adhesive.
- Art Fry had an idea to use microspheres to make sticky bookmarks.
- People started writing messages on Post-it Notes.
- 3M sent out samples of Post-it Notes to try to get people to buy them.

8

Forks Are Finally Accepted

People have been eating with spoons and knives since prehistoric times. But the fork is a newer invention. At first, forks were large and had two tines, or spikes. Cooks used them to spear large pieces of meat.

In 1004, smaller forks appeared at a wedding feast in Venice, Italy. The bride used them to eat candied fruits. Back then, using a fork was considered shameful. The Church felt that God had given people fingers to eat with. Finally, in the 1600s, King Charles of England declared it was decent to use a fork.

As the fork grew more common, its form changed. A third tine was added to make food less likely to slip through. A slight curve helped with scooping. By the 1700s, people brought their own forks to celebrations.

By the 1840s, utensils were less expensive. Ordinary people could afford to use forks. Most people enjoyed eating with a fork. Others, however, were not happy. One person complained that eating peas with a fork was like trying to eat soup with a knitting needle.

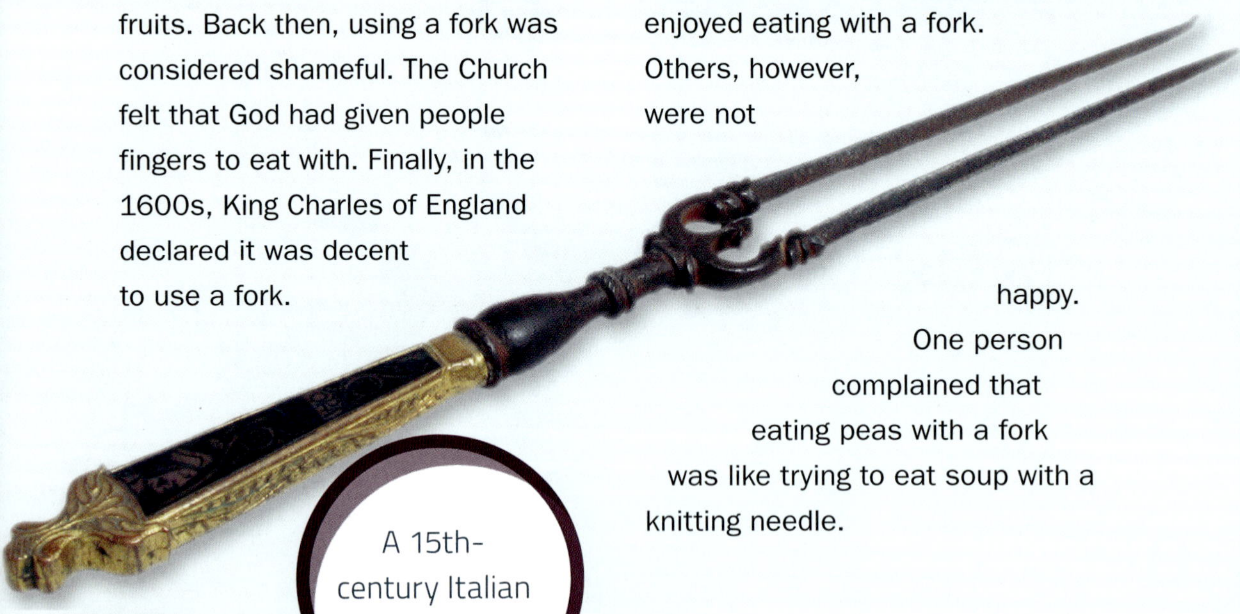

A 15th-century Italian fork.

2

Number of tines the first fork contained.

- Two-tined forks were first used to spear large pieces of meat.
- In 1004, forks were used at a wedding feast to eat fruit.
- By the 1800s, forks had 3 tines and a curved shape.

TOO MANY FORKS

As the fork became more popular, people started creating different forks for all types of food. There were forks for fish, pickles, salad, oysters, even ice cream. This created a problem for diners. Few knew the rules for the proper use of each fork. Finally, in 1935, US President Herbert Hoover made a law to limit the number of flatware pieces. Today most place settings only use two forks. One is for the salad and one is for the main course.

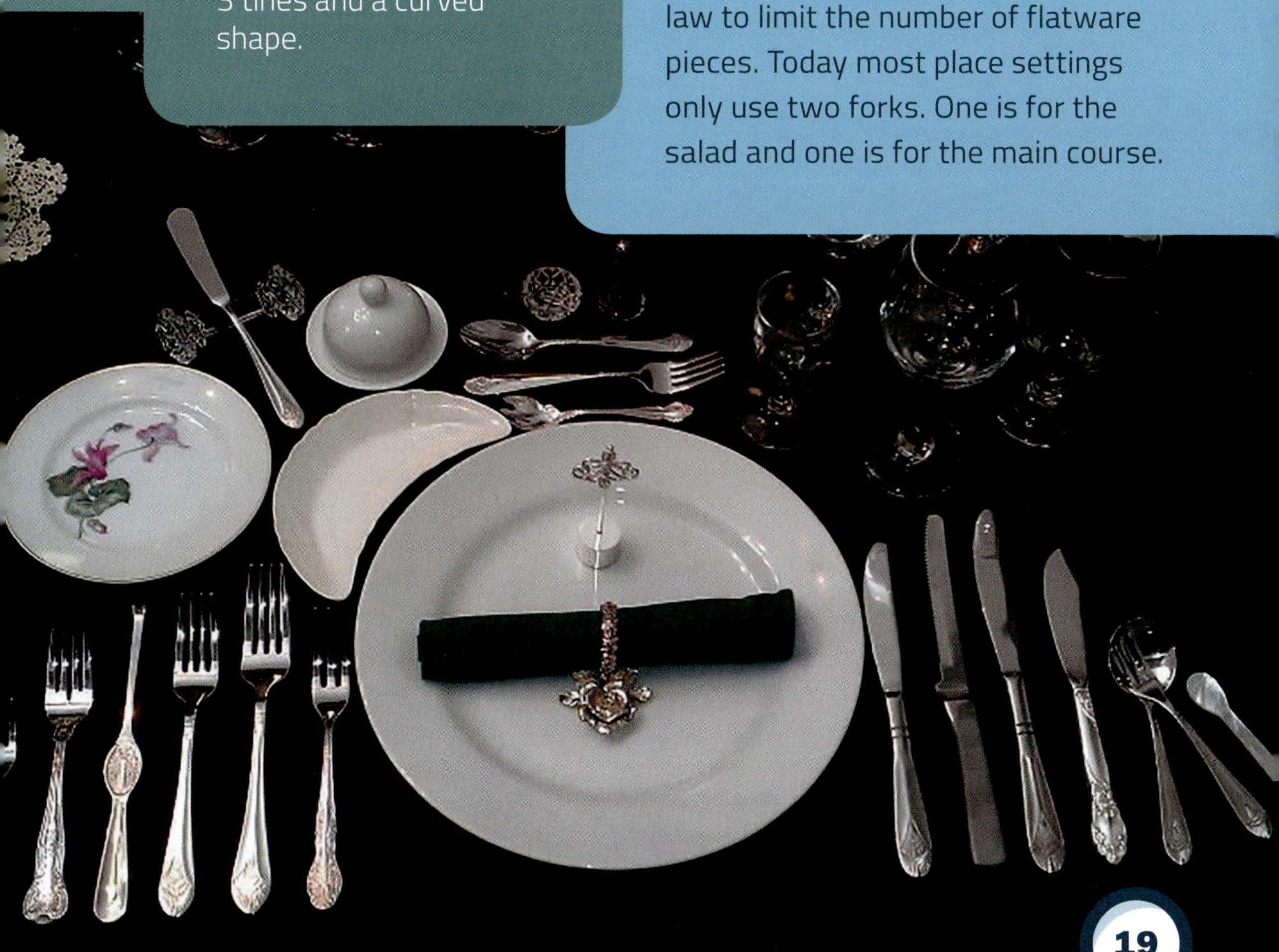

9

Burrs Inspire Velcro

After taking a walk in the woods one day, George de Mestral had to remove burrs from his pants. He wondered if the pesky prickles could ever be put to good use. After researching for years, de Mestral produced a new fastener. It was made of two strips of cloth. One strip had thousands of tiny hooks. The other had thousands of tiny loops. He named his invention Velcro. The name was a combination of the words velvet and crochet.

8

Years of research it took to invent Velcro.

- George de Mestral was inspired to make Velcro when picking burrs off his pants.
- Velcro is made of thousands of loops and hooks.
- NASA began to use Velcro in spacecraft in the 1960s.
- Puma made the first sneaker with Velcro fasteners.

Dinner in space involves using Velcro to secure everything to the table.

De DeMestral received a patent for his invention in 1955. Four years later, Velcro was introduced at a fashion show in New York City. News reporters declared it would put an end to buttons, hooks, snaps, and zippers.

NASA gave Velcro a big boost in the early 1960s. Astronauts used it to secure objects they didn't want floating around their spacecrafts. Soon Velcro was everywhere. Nurses used it to attach blood pressure gauges to people's arms. It kept hospital gowns closed. Cars used Velcro to keep floor mats from slipping. Airplanes used it to fasten seat cushions.

In 1968, Puma was the first shoe company to use Velcro on sneakers. By the end of the 1980s, it was widely used on children's shoes. Fastening three Velcro straps was much easier than tying a shoelace. Today the fastener is used on clothing, toys, and sports equipment. Almost anything that needs to stick together can be attached with Velcro.

10

Zippers Make Dressing Easier

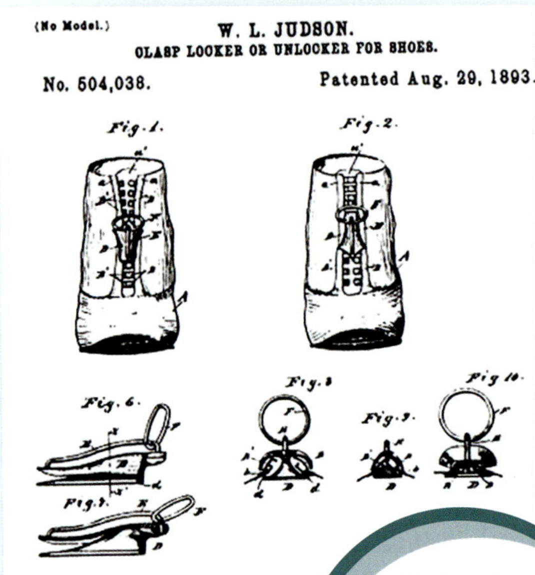

Judson's patent for the "Clasp Locker or Unlocker for Shoes."

In the late 1890s, getting dressed was a struggle. Women had to fasten buttons from their necklines to their feet. Boots alone had 20 to 30 buttons. To solve this problem, Whitcomb Judson invented the "Clasp Locker or Unlocker for Shoes." His hook-and-eye closure fastened boots in a single motion. He introduced his invention at the Chicago World's Fair in 1893. But it kept snapping open. Even though his fastener didn't work, Judson is known as the father of the zipper.

Gideon Sundback worked for years to improve on Judson's design. Finally, in 1914, he got rid of the hooks. Instead, he used two rows of tiny metal teeth, shaped like spoons. Each piece twisted and stayed locked to its mate on the other

10

Number of teeth per inch in the first zipper.

- Zippers were invented to make dressing easier.
- The first zipper didn't work well.
- Gideon Sundback perfected the design for the zipper used today.

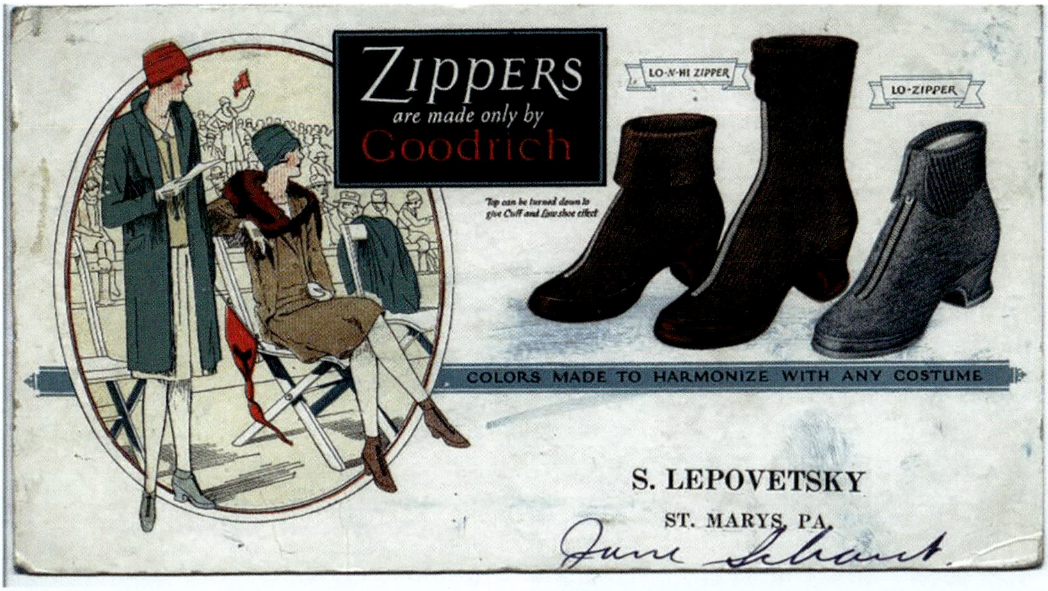

row. Sundback named his invention the "Hookless Fastener No. 2."

Zippers didn't catch on until 1923. B.F. Goodrich used them on galoshes. The rubber company gave the fastener its nickname. Zipper was short for "Zip 'er up." It was a play on the "zip" sound made by the metal teeth.

The invention spread throughout the clothing industry. Children's clothes were the first to have zippers. Soon zippers replaced button flies on pants. Today the fasteners come in many colors. They are made of plastic and nylon, as well as metal. Zippers are now found on everything from luggage to tent flaps.

WHY YKK?

The YKK printed on most of the metal pulltabs on today's zippers stands for Yoshida Kogyo Kabushikikaisha. This is the name of a Japanese manufacturer. They make about half of all the zippers in the world.

11 Better, Not Bitter, Peanut Butter

The ancient Incas and Aztecs were the first people to grind roasted peanuts into a paste. The creation was not very smooth, and it had a bitter taste. Several men deserve credit for the peanut butter eaten today. Marcellus Gilmore Edson of Canada patented peanut paste in 1884. It was made by milling roasted peanuts between heated surfaces. The next year, Dr. John Kellogg patented a process for making peanut

3

Pounds of peanut butter the average American eats each year.

- Ancient Aztecs and Incas ground roasted peanuts into a paste.
- Several inventors made peanut butter from raw or roasted peanuts.
- George Washington Carver made tastier peanut butter by adding sugar and molasses.

A 1950 Peter Pan advertisement from Life magazine.

The Jif peanut butter production plant in Kentucky.

butter from raw peanuts. He marketed it to his older patients as a protein-packed substitute for meat. Neither of these pastes was very tasty, however.

George Washington Carver found a way to make a tastier peanut butter. In 1903, he added sugar and molasses to an existing formula. The new creation was sweeter and had a slightly smoother texture. However, Carver's peanut butter was still a bit gritty.

Finally, in the early 1920s, Joseph Rosenfield began churning peanuts like butter. He made peanut butter that was smooth and spreadable. He licensed his process to the Pond Company, which began making Peter Pan peanut butter. In 1930, Rosenfield added chopped peanuts to the formula. He called his brand Skippy. It was the first crunchy style peanut butter.

Today Jif has the largest peanut butter plant in the world. They make 250,000 jars every day. About half of the peanuts grown in the United States are used for making peanut butter. The popular sandwich spread is both nutritious and delicious.

12 Ruth Wakefield Creates the Chocolate Chip Cookie

Ruth Wakefield liked to experiment with food. Her guests at the Toll House Inn in Whitman, Massachusetts, loved her thin butterscotch cookies. But she wanted to give them something different. In 1938, while returning from a trip to Egypt, Ruth worked on a new recipe.

Back in her kitchen, she took an ice pick and chopped up a Nestlé's chocolate bar. She added the

Ruth Wakefield and the Toll House Inn in Wakefield, Massachusetts in the 1930s.

Before the 1930s, no one had ever eaten a chocolate chip cookie. Cookies had been around since colonial times. But no one had ever put candy bits in the dough. Chocolate was used in cookies, but it was always melted first.

CHIPS VS. MORSELS

Nestlé wanted to make Toll House Cookies easier to prepare. In 1939, the company began to score their chocolate bars so they were easier to break into pieces. Soon after, the teardrop-shaped pieces were introduced. Nestlé calls them morsels. Other companies call them chips.

6

Billions of packages of chocolate chip cookies sold in the US each year.

- Before the 1930s, chocolate was melted before it was added to a cookie recipe.
- In 1938, Ruth Wakefield added a chopped-up candy bar to her cookie dough.
- Wakefield's Toll House Crunch cookies became popular.
- Nestlé bought the recipe for Toll House Cookies from Ruth Wakefield for $1.

chunks to the butterscotch cookie recipe. She called her creation the Toll House Crunch Cookie. It was small and crisp.

Ruth published the recipe in a Boston newspaper. Sales of Nestlé chocolate bars spiked. Andrew Nestlé bought the recipe for one dollar. It was printed on every package of the chocolate.

Love of Toll House cookies spread throughout the country. Today millions of chocolate chip cookies are made in factories. Half of all the cookies baked at home are chocolate chip cookies. Chocolate chip is America's favorite cookie.

Fact Sheet

- The Frisbee has inspired several sports that use the flying disc. High school students invented Ultimate Frisbee in 1967. The game is a cross between football, soccer, and basketball. There are now hundreds of courses for Frisbee Golf, where the discs are tossed into metal baskets. In Freestyle Frisbee, discs fly during choreographed routines set to music. Players create their own moves and styles.

- Microwave ovens use high-frequency radio waves to cook food. Water molecules inside the food absorb the microwaves. They start to vibrate. This creates the heat that cooks the food. Microwaves bounce off metal. This is why manufacturers say not to use metal containers or aluminum foil in a microwave oven. Your food won't cook. And microwaves bouncing around inside the oven can cause sparks, even fires.

- George Washington Carver didn't invent peanut butter. But he discovered more than 300 uses for the tasty spread. He even used it in things people didn't eat. Some of these include glue, shaving cream, and shampoo. Today peanut butter is found in 94 percent of kitchens in America.

George Washington Carver in 1906.

- Forks became popular for several reasons. Eating with a fork is more sanitary than eating with bare hands. It's also a safer way to eat meat than by chewing it off the tip of a sharp knife. Before forks, people usually wiped their hands on the tablecloth when they were done eating. It's easier and cheaper to wash forks than to clean stained linens.

- During the 1930s, Americans were suffering from the Great Depression. Ruth Wakefield's cookie creation was a perfect escape, if only for a moment. Biting into a warm chocolate chip cookie gave comfort to millions of people. When the United States entered World War II, the comfort spread overseas. Many wives and mothers packed Toll House cookies into care packages they shipped to their loved ones fighting in other countries.

Glossary

closure
Something that closes an opening.

commercial
Objects or machines used in business or industry.

consumers
Buyers of goods and services.

galoshes
Rubber boots made to wear over regular shoes to keep water out.

gauges
Instruments used for measuring.

infrared
Light waves that can't be seen by the human eye.

microspheres
Very tiny, hollow particles that can only be seen with a microscope.

millennium
A period of 1,000 years.

patent
An official document that gives a person or company who invents or discovers something the right to be the only one who makes or sells it for a certain period of time.

prototype
A full-size model of an item that will be manufactured.

receptors
Devices that receive electronic signals.

utensils
Tools for eating or cooking.

For More Information

Books

Bender, Lionel, and Dave King. *Invention*. New York: DK, 2013.

Maggs, Sam, and Sophia Foster-Dimino. *Wonder Women: 25 Innovators, Inventors, and Trailblazers Who Changed History*. Philadelphia, PA: Quirk Books, 2016.

Turner, Tracey, et al. *100 Inventions That Made History: Brilliant Breakthroughs That Shaped Our World*. New York: DK Publishing, 2016.

Williams, Marcia. *Hooray for Inventors!* Somerville, MA: Candlewick Press, 2013.

Index

About the Author

Patricia Hutchison is a former teacher. She has written numerous nonfiction books about history, science, and nature. When she is not writing, she loves to travel with her husband and family.